~A BINGO BOOK~

Genetics and Heredity Bingo Book

COMPLETE BINGO GAME IN A BOOK

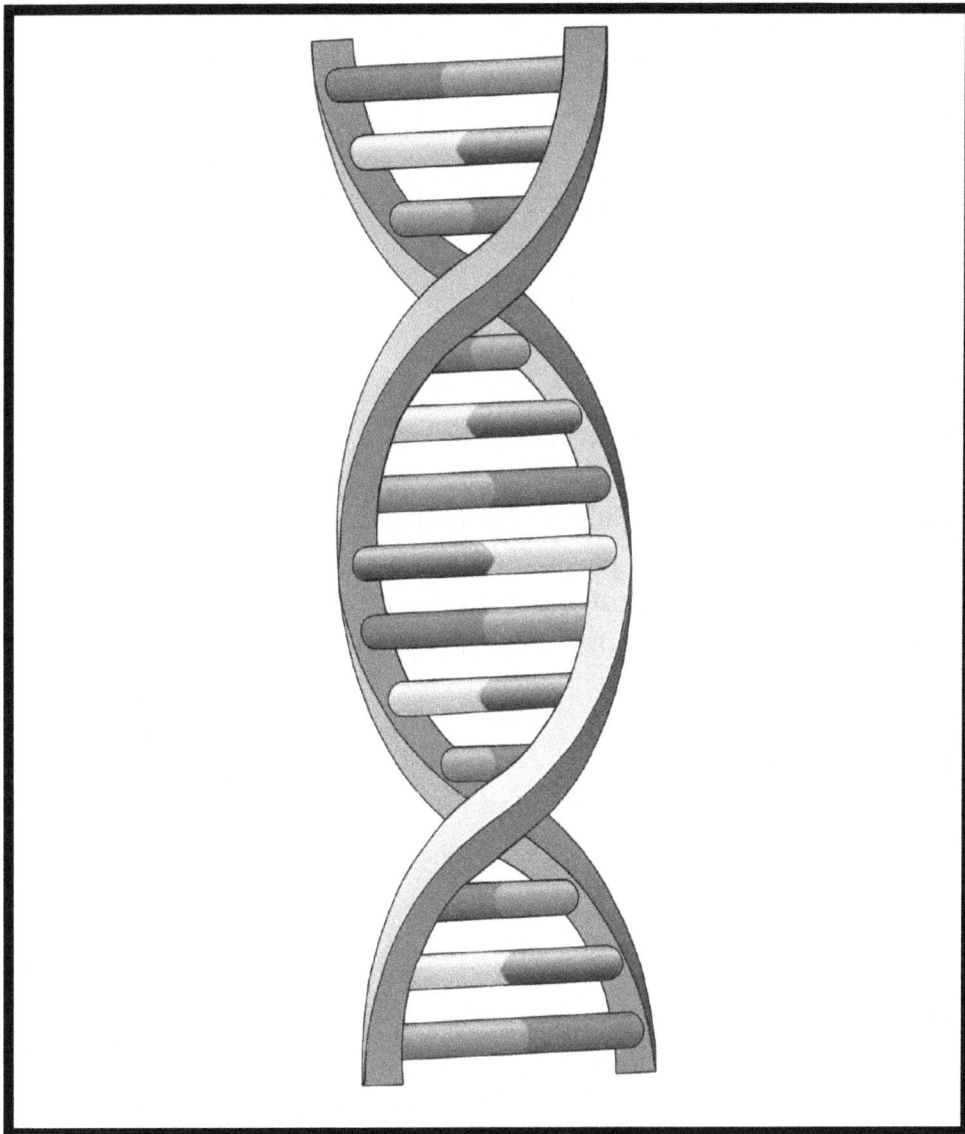

I0000348

Written By Rebecca Stark
Educational Books 'n' Bingo

© 2016 Barbara M. Peller, also known as Rebecca Stark

The purchase of this book entitles the buyer to exclusive reproduction rights of the student activity pages for his or her class only. The reproduction of any part of the work for an entire school or school system or for commercial use is prohibited.

Educational Books 'n' Bingo

ISBN 978-0-87386-450-3

Printed in the U.S.A.

GENETICS & HEREDITY BINGO DIRECTIONS

INCLUDED:

List of Terms

Templates for Additional Terms and Clues

2 Clues per Term

30 Unique Bingo Cards

Markers

1. **Either cut apart the book or make copies of ALL the sheets. You might want to make an extra copy of the clue sheets to use for introduction and review. Keep the sheets in an envelope for easy reuse.**

2. Cut apart the call cards with terms and clues.

3. Pass out one bingo card per student. There are enough for a class of 30.

4. Pass out markers. You may cut apart the markers included in this book or use any other small items of your choice.

5. Decide whether or not you will require the entire card to be filled. Requiring the entire card to be filled provides a better review. However, if you have a short time to fill, you may prefer to have them do the just the border or some other format. Tell the class before you begin what is required.

6. There are 50 terms. Read the list before you begin. If there are any terms that have not been covered in class, you may want to read to the students the term and clues before you begin.

7. There is a blank space in the middle of each card. You can instruct the students to use it as a free space or you can write in answers to cover terms not included. Of course, in this case you would create your own clues. (Templates provided.)

8. Shuffle the cards and place them in a pile. Two or three clues are provided for each term. If you plan to play the game with the same group more than once, you might want to choose a different clue for each game. If not, you may choose to use more than one clue.

9. Be sure to keep the cards you have used for the present game in a separate pile. When a student calls, "Bingo," he or she will have to verify that the correct answers are on his or her card AND that the markers were placed in response to the proper questions. Pull out the cards that are on the student's card keeping them in the order they were used in the game. Read each clue as it was given and ask the student to identify the correct answer from his or her card.

10. If the student has the correct answers on the card AND has shown that they were marked in response to the *correct questions,* then that student is the winner and the game is over. If the student does not have the correct answers on the card OR he or she marked the answers in response to *the wrong questions,* then the game continues until there is a proper winner.

11. If you want to play again, reshuffle the cards and begin again.

Have fun!

© **Barbara M. Peller**

TERMS INCLUDED

allele(s)	hereditary
biology	heredity
blood type(s)	heterozygous
cell	homozygous
chromosome(s)	inherit(ed)
clone(d)	meiosis
codominance	Mendel
Crick and Watson	mitosis
cytoplasm	mutation
Darwin	Nobel Prize
diploid	nucleic acids
DNA	nucleotides
dominant	nucleus
double helix	offspring
evolution	pedigree
engineering	phenotype
fruit fly	probability
gamete	proteins
gene(s)	Punnett square
geneticist	recombinant
genetic disorder	recessive
genetics	RNA
genome	traits
genotype	twins
hemophilia	zygote

© Barbara M. Peller

Additional Terms

Choose as many other terms as you would like and write them in the squares.
Repeat each as desired.
Cut out the squares and randomly distribute them to the class.
Instruct the students to place the square on the center space of their card.

© Barbara M. Peller

Clues for Additional Terms

Write two or three clues for each of your additional terms.

_____ 1. 2. 3.	_____ 1. 2. 3.
_____ 1. 2. 3.	_____ 1. 2. 3.
_____ 1. 2. 3.	_____ 1. 2. 3.

Genetics and Heredity Bingo

© Barbara M. Peller

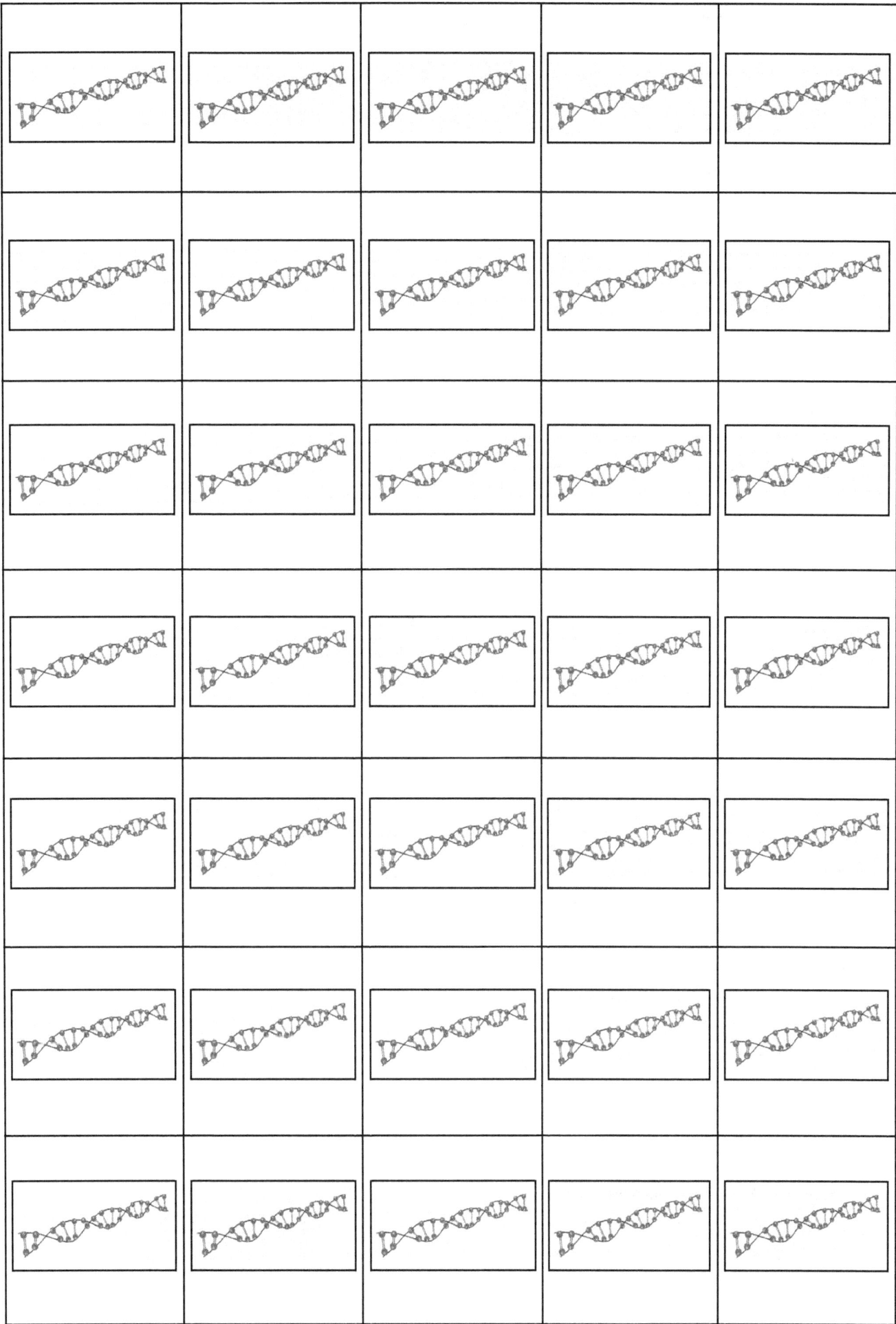

allele(s) 1. It is one of a pair or series of genes that occupy a specific position on a specific chromosome. 2. ___ are different forms of the same gene.	**biology** 1. Genetics is a branch of ___. 2. Genetics is the branch of ___ that studies heredity.
blood type(s) 1. ABO is a classification system for ___. 2. If the mother has an O allele and the father has an A allele, the offspring's ___ will be A.	**cell** 1. It is larger than a chromosome and smaller than a tissue. 2. Chromosomes are found in the nucleus of a ___.
chromosome(s) 1. Most cells in the human body have 23 pairs of ___. 2. Normal females have two X ___. Normal males have both an X ___ and a Y ___.	**clone(d)** 1. It is an organism that is genetically identical to the unit or individual from which it was derived. 2. A sheep named Dolly became the first mammal to be ___ from an adult somatic cell.
codominance 1. When neither allele of a gene pair in a heterozygote is dominant or recessive to the other, this condition exists. 2. Pink snapdragons develop from a red and white cross because of ___.	**Crick and Watson** 1. They proposed the double helix structure of DNA. 2. They won the Nobel Prize in Physiology or Medicine in 1962.
cytoplasm 1. In eukaryotic cells, the cytoplasm is that part of the cell between the cell membrane and the nuclear envelope. 2. RNA is found in both the nucleus and the ___ of the cell.	**Darwin** 1. This British naturalist established that all species of life descended from common ancestry. 2. He is best known for his theory of evolution.

Genetics and Heredity Bingo

© Barbara M. Peller

diploid	**DNA**
1. A cell or organism that has a pair of each type of chromosome is called ___. 2. Most ___ organisms have one set of chromosomes from the mother and one set from the father.	1. It stands for deoxyribonucleic acid. 2. It is the hereditary material in most organisms.
dominant	**double helix**
1. If one gene for brown eyes and one gene for blue eyes are present, the person will have brown eyes because the gene for brown eyes is ___. 2. When expressing ___ and recessive alleles, the dominant allele is capitalized.	1. This describes the shape of a molecule of DNA. 2. James Watson and Francis Crick proposed the ___ structure of DNA.
evolution	**engineering**
1. ___ is a process that results in heritable changes in a population spread over many generations. 2. Biological ___ refers to changes in populations over generations and not to individuals.	1. Genetic ___, or genetic modification, is the human manipulation of an organism's genetic material. 2. Genetic ___ is the alteration of genetic code by artificial means.
fruit fly	**gamete**
1. *Drosophila melanogaster,* or ___ , is a good research animal because it develops from fertilized egg to embryo within 9 days. 2. In 1995 Drs. Lewis, Nuesslein-Volhard, and Wieschaus were awarded the Nobel Prize in Medicine because of their research with this animal.	1. A ___ is a mature sexual reproductive cell. 2. It unites with another ___ to form a new organism.
gene(s)	**geneticist**
1. A ___ is a unit of heredity. 2. Our ___ determine particular hereditary traits.	1. A scientist who studies genetics is called a ___. 2. A ___ is a type of biologist.

Genetics and Heredity Bingo

© **Barbara M. Peller**

genetic disorder	**genetics**
1. Sickle-cell anemia is one; it is a blood disease. 2. Tay-Sachs disease is one; it is a metabolic disorder that affects the brain and other tissues.	1. It is the branch of biology that studies heredity. 2. It is the study of genes and their relation to heredity.
genome	**genotype**
1. The set of genes a species has on all its chromosomes is called a ___. 2. Scientists still do not know how many genes are in the human ___.	1. This is the inheritable, internally coded information carried by all living organisms. 2. The ___ provides codes for the phenotype.
hemophilia	**hereditary**
1. If a person has this inherited blood disorder, blood does not coagulate properly. 2. This is a hereditary blood-coagulation disorder.	1. Something that is genetically transmitted from parent to offspring is said to be ___. 2. Inherited genetic characteristics are said to be ___.
heredity	**heterozygous**
1. Genetics is the branch of biology that studies genes and ___. 2. Complete this analogy: botany : plants :: genetics : ___	1. ___ refers to having two different alleles for the same trait. 2. If an organism has one dominant allele and one recessive allele for a particular trait, it is ___ for that trait.
homozygous	**inherit(ed)**
1. If an organism has two identical alleles for a particular trait, it is ___ for that trait. 2. Complete this analogy: heterozygous : different :: ___ : same	1. To receive a characteristic from one's parents by genetic transmission is to ___ that characteristic. 2. If you and your mother have the same blue eyes, you ___ your blue eyes from your mother.

© **Barbara M. Peller**

meiosis	**Mendel**
1. ___ is the process in cell division that reduces the number of chromosomes to half the original number.	1. Gregor ___ is often called the Father of Genetics.
2. In ___ the number of chromosomes are reduced from diploid to haploid.	2. His experiments showed that the inheritance of certain traits in pea plants follow certain laws.

mitosis	**mutation**
1. In this process of cell division two daughter cells are produced from a single parent cell.	1. A relatively permanent change in hereditary material is called a ___.
2. In this process of cell division the nucleus divides into nuclei containing the same number of chromosomes as the parent nucleus.	2. A change of the DNA sequence within a gene or chromosome resulting in the creation of a new trait not in the parental type is a ___.

Nobel Prize	**nucleic acids**
1. Watson and Crick won the ___ in Physiology or Medicine in 1962.	1. DNA and RNA are ___.
2. In 2007 Mario R. Capecchi, Martin J. Evans and Oliver Smithies received this for their work in stem-cell research.	2. ___ are biological molecules essential for life. They work together to help cells replicate and build proteins.

nucleotides	**nucleus**
1. Nucleic acids are made up of building blocks called ___.	1. DNA is found in the ___ of a cell.
2. The 5 ___ that make up nucleic acids are uracil, cytosine, thymine, adenine, and guanine.	2. A eukaryote is an organism with a membrane-enclosed ___.

offspring	**pedigree**
1. New organisms produced by a living organism are its ___.	1. A chart of an individual's ancestors used to analyze the inheritance of certain traits is called a ___.
2. The product of the reproductive processes of an animal or plant is the organism's ___.	2. This type of diagram is often used to analyze the inheritance of familial diseases across several generations.

© **Barbara M. Peller**

phenotype

1. It is the entire physical, biochemical, and physiological makeup of an individual, determined both genetically and environmentally.

2. The ___ is the outward, physical manifestation of the organism.

probability

1. The expected frequency of a particular event is the ___.

2. If one parent has a gene for brown eyes and a gene for blue eyes and the other has 2 genes for blue eyes, there is a 50% ___ that their child will have brown eyes.

proteins

1. Each cell has thousands of ___, each with a different function.

2. Amino acids are the building blocks of ___.

Punnett square

1. This diagram is used to predict an outcome of a particular cross or breeding experiment.

2. A ___ is a chart used to show the possible combinations of alleles in a cross of parents with known genotypes.

recombinant

1. ___ DNA is a form of artificial DNA that is created by combining two or more sequences that would not normally occur together.

2. rDNA, or ___ DNA, is the taking a strand of one DNA and combining it with another.

recessive

1. An allele that does not produce a characteristic effect when present with a dominant allele is said to be ___.

2. Complete this analogy:

dominant : ___ :: controlling : submissive

RNA

1. It stands for ribonucleic acid. It is found in the nucleus and the cytoplasm of the cell.

2. Its main function is to transfer the genetic code needed for the creation of proteins from the nucleus to the ribosome.

traits

1. They are the characteristics of an organism.

2. Some ___ are expressed by genes; others are influenced by the environment.

twins

1. Monozygotic ___ are genetically identical.

2. Monozygotic, or identical, ___ are always the same sex.

zygote

1. A ___ is a single cell that contains the genetic material of both the mother and the father.

2. When a cell begins to divide, it is no longer called a ___; it is called an embryo.

Genetics and Heredity Bingo

© Barbara M. Peller

Genetics and Heredity Bingo

genome	genetic disorder	genotype	zygote	RNA
Crick and Watson	allele(s)	twins	inherit(ed)	genetics
probability	nucleotides		hemophilia	mutation
recessive	cell	gene(s)	Punnett square	hereditary
heredity	cytoplasm	DNA	gamete	geneticist

Genetics and Heredity Bingo: Card No. 1

© Barbara M. Peller

Genetics and Heredity Bingo

recessive	proteins	heterozygous	nucleic acids	heredity
hereditary	inherit(ed)	clone(d)	cell	phenotype
offspring	cytoplasm		dominant	gene(s)
fruit fly	pedigree	nucleotides	Mendel	genetics
geneticist	twins	DNA	Crick and Watson	gamete

© Barbara M. Peller

Genetics and Heredity Bingo

recessive	gene(s)	inherit(ed)	Punnett square	probability
cytoplasm	allele(s)	blood type(s)	genetic disorder	meiosis
cell	twins		phenotype	biology
nucleotides	offspring	heredity	fruit fly	heterozygous
gamete	Crick and Watson	DNA	Mendel	genotype

© Barbara M. Peller

Genetics and Heredity Bingo

nucleotides	phenotype	heredity	Crick and Watson	genotype
homozygous	clone(d)	genetic disorder	nucleic acids	probability
hemophilia	fruit fly		RNA	zygote
gene(s)	engineering	twins	DNA	blood type(s)
evolution	geneticist	Nobel Prize	gamete	mutation

© Barbara M. Peller

Genetics and Heredity Bingo

geneticist	RNA	cell	clone(d)	Crick and Watson
homozygous	gene(s)	blood type(s)	dominant	allele(s)
proteins	mutation		double helix	diploid
genetics	phenotype	genome	Mendel	evolution
inherit(ed)	DNA	nucleus	nucleotides	hemophilia

© Barbara M. Peller

Genetics and Heredity Bingo

biology	phenotype	heterozygous	proteins	mutation
Punnett square	cell	evolution	genetic disorder	probability
nucleic acids	blood type(s)		clone(d)	dominant
DNA	heredity	Mendel	Nobel Prize	hemophilia
hereditary	gene(s)	genome	nucleus	genotype

Genetics and Heredity Bingo: Card No. 6

© Barbara M. Peller

Genetics and Heredity Bingo

genome	phenotype	diploid	double helix	inherit(ed)
hereditary	genotype	cytoplasm	allele(s)	homozygous
heterozygous	zygote		dominant	Darwin
nucleotides	fruit fly	probability	recessive	offspring
DNA	Crick and Watson	Mendel	Nobel Prize	biology

© Barbara M. Peller

Genetics and Heredity Bingo

hemophilia	phenotype	chromo-some(s)"	Punnett square	Darwin
homozygous	proteins	nucleic acids	mutation	clone(d)
probability	mitosis		genotype	RNA
gamete	nucleotides	recessive	evolution	fruit fly
twins	DNA	Nobel Prize	cell	hereditary

© Barbara M. Peller

Genetics and Heredity Bingo

dominant	inherit(ed)	cytoplasm	probability	mutation
evolution	proteins	hemophilia	cell	genotype
meiosis	genome		allele(s)	chromosome(s)
Darwin	geneticist	heredity	double helix	diploid
fruit fly	Mendel	blood type(s)	recessive	RNA

© Barbara M. Peller

Genetics and Heredity Bingo

recessive	Punnett square	clone(d)	nucleic acids	nucleus
mutation	Darwin	genetic disorder	allele(s)	genotype
mitosis	phenotype		zygote	offspring
heredity	genetics	evolution	Mendel	meiosis
codominance	hereditary	heterozygous	geneticist	hemophilia

© Barbara M. Peller

Genetics and Heredity Bingo

biology	phenotype	cell	evolution	hereditary
chromosome(s)	meiosis	double helix	dominant	genetic disorder
homozygous	proteins		heterozygous	cytoplasm
codominance	probability	Mendel	Crick and Watson	recessive
blood type(s)	DNA	genome	Nobel Prize	inherit(ed)

© Barbara M. Peller

Genetics and Heredity Bingo

inherit(ed)	RNA	meiosis	Punnett square	dominant
cytoplasm	twins	proteins	Nobel Prize	allele(s)
genome	diploid		mutation	nucleic acids
DNA	fruit fly	genotype	recessive	homozygous
phenotype	chromosome(s)	mitosis	blood type(s)	Darwin

© Barbara M. Peller

Genetics and Heredity Bingo

codominance	RNA	biology	meiosis	mutation
proteins	chromosome(s)	phenotype	dominant	offspring
Punnett square	clone(d)		cytoplasm	diploid
hemophilia	Mendel	Darwin	mitosis	recessive
DNA	genetics	Nobel Prize	genome	double helix

                    © Barbara M. Peller

Genetics and Heredity Bingo

Crick and Watson	proteins	cell	dominant	codominance
Darwin	genome	meiosis	allele(s)	phenotype
evolution	zygote		heterozygous	blood type(s)
genetics	Mendel	mitosis	clone(d)	biology
DNA	nucleic acids	offspring	hereditary	hemophilia

© Barbara M. Peller

Genetics and Heredity Bingo

double helix	dominant	cell	inherit(ed)	Punnett square
biology	heterozygous	genetic disorder	proteins	evolution
mutation	genome		probability	genotype
DNA	meiosis	chromosome(s)	Mendel	codominance
hereditary	fruit fly	Nobel Prize	nucleus	cytoplasm

Genetics and Heredity Bingo: Card No. 15

© Barbara M. Peller

Genetics and Heredity Bingo

clone(d)	meiosis	chromosome(s)	nucleus	pedigree
nucleic acids	offspring	diploid	homozygous	zygote
codominance	RNA		mutation	cytoplasm
nucleotides	Darwin	DNA	double helix	recessive
evolution	traits	Nobel Prize	fruit fly	phenotype

Genetics and Heredity Bingo: Card No. 16

© Barbara M. Peller

Genetics and Heredity Bingo

codominance	recombinant	engineering	meiosis	Crick and Watson
double helix	evolution	Mendel	zygote	diploid
dominant	recessive		traits	chromosome(s)
geneticist	hereditary	hemophilia	cell	offspring
heredity	blood type(s)	inherit(ed)	Punnett square	RNA

© Barbara M. Peller

Genetics and Heredity Bingo

genotype	mitosis	Darwin	evolution	nucleic acids
phenotype	codominance	heredity	mutation	blood type(s)
dominant	offspring		engineering	nucleus
geneticist	genetic disorder	Mendel	recessive	heterozygous
traits	meiosis	cell	recombinant	biology

Genetics and Heredity Bingo: Card No. 18

© Barbara M. Peller

Genetics and Heredity Bingo

mutation	biology	meiosis	chromosome(s)	mitosis
double helix	Punnett square	nucleus	inherit(ed)	zygote
recombinant	Crick and Watson		allele(s)	genotype
heterozygous	traits	heredity	fruit fly	engineering
probability	pedigree	hereditary	hemophilia	Nobel Prize

© Barbara M. Peller

Genetics and Heredity Bingo

mitosis	recombinant	Punnett square	meiosis	allele(s)
clone(d)	cytoplasm	homozygous	heredity	nucleic acids
RNA	diploid		nucleotides	genetic disorder
geneticist	hemophilia	gamete	fruit fly	traits
gene(s)	twins	pedigree	recessive	engineering

© Barbara M. Peller

Genetics and Heredity Bingo

double helix	biology	homozygous	meiosis	genetics
RNA	engineering	Darwin	chromosome(s)	genome
offspring	hereditary		recombinant	cell
heredity	inherit(ed)	traits	geneticist	hemophilia
nucleotides	pedigree	Nobel Prize	codominance	fruit fly

© Barbara M. Peller

Genetics and Heredity Bingo

probability	heterozygous	engineering	proteins	codominance
nucleic acids	Punnett square	genotype	chromosome(s)	allele(s)
Darwin	zygote		genome	diploid
traits	geneticist	fruit fly	genetic disorder	Crick and Watson
pedigree	blood type(s)	recombinant	offspring	homozygous

© Barbara M. Peller

Genetics and Heredity Bingo

clone(d)	recombinant	inherit(ed)	proteins	Nobel Prize
biology	mitosis	hereditary	double helix	genetic disorder
heterozygous	codominance		gamete	genome
offspring	pedigree	traits	blood type(s)	fruit fly
genetics	hemophilia	twins	heredity	engineering

© Barbara M. Peller

Genetics and Heredity Bingo

clone(d)	mitosis	Crick and Watson	recombinant	chromosome(s)
mutation	Nobel Prize	homozygous	nucleic acids	genome
diploid	nucleus		codominance	offspring
genetics	gamete	traits	blood type(s)	RNA
gene(s)	nucleotides	pedigree	Punnett square	twins

Genetics and Heredity Bingo: Card No. 24

© Barbara M. Peller

Genetics and Heredity Bingo

nucleotides	homozygous	recombinant	cell	engineering
genetic disorder	genetics	double helix	clone(d)	allele(s)
RNA	chromosome(s)		gamete	traits
nucleus	geneticist	twins	pedigree	zygote
Nobel Prize	Crick and Watson	Darwin	evolution	gene(s)

© Barbara M. Peller

Genetics and Heredity Bingo

engineering	recombinant	gamete	nucleic acids	nucleus
heredity	Punnett square	chromosome(s)	mitosis	clone(d)
genetics	heterozygous		zygote	nucleotides
codominance	proteins	geneticist	pedigree	traits
diploid	evolution	cell	twins	gene(s)

© Barbara M. Peller

Genetics and Heredity Bingo

gamete	Darwin	recombinant	mitosis	cytoplasm
genetics	heterozygous	double helix	traits	allele(s)
Mendel	twins		pedigree	nucleotides
nucleus	biology	homozygous	gene(s)	genetic disorder
codominance	zygote	engineering	probability	diploid

© Barbara M. Peller

Genetics and Heredity Bingo

mutation	mitosis	recessive	recombinant	Darwin
cytoplasm	engineering	gamete	heredity	zygote
twins	offspring		nucleus	nucleic acids
diploid	probability	hereditary	pedigree	traits
proteins	dominant	codominance	gene(s)	genetics

© Barbara M. Peller

Genetics and Heredity Bingo

engineering	mitosis	nucleus	double helix	dominant
genetics	heredity	homozygous	diploid	probability
RNA	gamete		allele(s)	recombinant
cytoplasm	geneticist	genotype	pedigree	traits
clone(d)	chromosome(s)	gene(s)	biology	twins

© Barbara M. Peller

Genetics and Heredity Bingo

Crick and Watson	recombinant	nucleic acids	dominant	RNA
genetic disorder	nucleus	heterozygous	zygote	allele(s)
gene(s)	blood type(s)		diploid	homozygous
genetics	biology	mitosis	pedigree	gamete
geneticist	inherit(ed)	twins	engineering	genotype

© Barbara M. Peller

www.ingramcontent.com/pod-product-compliance
Lightning Source LLC
Chambersburg PA
CBHW051428200326
41520CB00023B/7395